Cranberry

Voices from the Bogs

Memories

Edmund Lodi
Yolanda Lodi

Cranberry

Voices from the Bogs

Memories

compiled and edited

by

Edward and Yolanda Lodi

ROCK VILLAGE PUBLISHING

Middleborough, Massachusetts

First Printing

Cranberry Memories: Voices from the Bogs

Photographs: Front cover and half-title page by John B. Howes
Back cover illustration by John B. Howes

Typography and cover design by Carolyn Gilmore

ISBN 978-1-934400-17-3

Rock Village Publishing
41 Walnut Street
Middleborough MA 02346

(508) 946-4738

rockvillage@verizon.net

Contents

Cranberry

Voices from the Bogs

Memories

Preface

Those of us who live in or visit southeastern Massachusetts have some awareness of the economic value of cranberries. But what of the cultural significance of this all-American fruit? What of the many ways in which cranberry agronomy influences our lives—not only in respect to the physical environment, the ways the bogs with their swamps and uplands, their sand pits and reservoirs, sculpt the landscape that surrounds us, but in respect to our inner being as well, the ways that working on the bogs, or living near them and enjoying the natural world they promote, help mold our character and emotional health, our very outlook on life?

The son of one grower remembers his mother proudly proclaiming, "We're bog people," as if there could be no greater recommendation for her family's integrity. The daughter of another grower talks of her father's "untimely" death at the age of 95—untimely, because at that advanced age he was still going strong, and actually died while driving his pickup around his beloved bog.

Beloved—more than any other, that word sums up the attitudes and feelings of the vast majority of the voices included in this anthology toward the little red berry, and the bogs on which it grows. Even those handful of writers who do not proclaim to love the bogs, and who remember only the back-breaking toil, often for low pay, or the heartbreak of crops destroyed by frost, drought, hail, or insect infestation—seem nonetheless to have a deep and abiding respect for the bogs, and for the men and women who earn their livings from them.

When the editors of Rock Village Publishing originally conceived the idea for this collection of memoirs and anecdotes, it was intended as a celebration of the importance of cranberries in the many aspects and phases of our lives. As the varied and remarkable stories began to arrive in the mail, however, we quickly discovered that we were also putting together an informal history: of the bogs, and of the people who shaped them, and whose lives have been shaped by them.

Cranberry Memories: Voices from the Bogs captures, we believe, colorful and unique chunks of American history and lore, many of which might otherwise have gone unrecorded: forgotten, forever lost to posterity.

A Paean to the Bogs

by
Lynda Ames

Only a "bog-trotter" can understand what it is to be passionate about the first-hand experience of planting, weeding, sanding, and harvesting. I suppose we are the only ones, too, who love the smell of fermented berries and dried vines in the warm September sun and that swampy odor that peaks after dark.

Funny, isn't it? It's one of the hardest jobs anyone will ever do, but from year to year you forget the sore muscles, the blistered hands, the scorching heat, and cold damp mornings waiting for the vines to dry.

Our family all agrees on one thing, though. No gourmet restaurant could serve a meal that tastes as good as a packed lunch at mid-day, eaten on an upturned cranberry box. Life was good.

Ray's Ruminations

by

Leslie Ray Sears, III

Cranberries have been part of my family's livelihood since 1819. Captain Henry Hall first cultivated cranberries in Dennis in 1816; it didn't take long for word to get around town about domesticating and selling this sour wild berry.

In the fall of 1819, my great-great grandfather Elkanah Sears (1758-1836) with his son William (1808-1896) set out some cranberry vines at Flax Pond (now called Scargo Lake) in Dennis. Elkanah Howes Sears, William's son, was a cooper, making cranberry barrels and boxes in his workshop on School Street in East Dennis. Elkanah traveled from Cape Cod to Wisconsin to buy wood to make barrels. What a huge amount of effort considering that in 1868 a standard 100 lb. barrel of cranberries sold for 58 cents. By 1888 shipments of cranberries (scientific classification: *vaccinium macrocarpon*) from Cape Cod were 80,128 barrels and 13,463 boxes.

Cranberries must be in tune with the number 3. There are approximately 333 cranberries in a pound, 3,333 cranberries in one gallon of juice and 33,333 cranberries in a 100-pound barrel. Isn't it curious that another species of the vaccinium genus is the blueberry? Cranberries and blueberries are cousins! Picking high bush blueberries at cultivated sites on the Cape is another favorite past-time.

I have found many family deeds for bogs and sand pits. Adding sand to a bog increased production of the berries. When bogs were flooded to protect them in winter, sand could be easily spread on the ice. The sand sifted to the vines in the spring when ice melted. Searching the Barnstable County Register of Deeds website may enable you to find out about your family's vocation based on land use.

I imagine my grandpa Leslie (1892-1954), Elkanah's son, was involved in the family business of cultivating, harvesting and boxing the berries along with his three siblings, Lewis, Lelia and Billy. Elkanah's cousin Edmund lived next door and is listed in the 1920 census as a cranberry barrel manufacturer. I know that in the 1940's, my dad, Les Sears, and his buddy Bob Duke picked cranberries for Lou Crowell in late summer and early fall.

Mr. Crowell, Dad's fourth cousin, lived on Main Street in East Dennis across the street from our Grandpa Smalley and would come creeping across Main Street from "Moose House" in the early morning hours to wake them up for work because he knew they had stayed up very late the night before playing cards. Dad related how they almost died in the noonday sun, picking cranberries on their hands and knees for 35 cents a box until they were so hot they were ready to drop. Then Dad and Bob would jump into the pond for a swim before lunch to revive themselves.

Cranberry vines are about six inches tall and many feet long. Dad picked cranberries dry with a flat bottomed, rectangular scoop about eighteen inches long and twelve inches wide that looks like a many-toothed comb with a box on the end. The "teeth" of the comb rip the red berries from the vine.

There was always a cranberry scoop in our house, usually at the end of the couch, full of magazines. I am sure when my Dad looked at that scoop and remembered the blisters, sore knees and sunburns, that he was thankful that the GI Bill helped him through college. An antique scoop is a valuable commodity these days and in fact, when our house in East Dennis was burglarized, that scoop was one of the few items that were stolen.

During World War II, while Dad was in high school, some readers will remember rationing. Gasoline was rationed and, following the Japanese conquest of Malaya, so was rubber. (The U. S. Office of the Rubber Director was charged with getting used tires to factories, where they became parts for tanks and trucks.)

Due to rubber rationing, the tire tread was worn so thin that flat tires were common. Dad bragged how they had the tire changing process down to a number of minutes and usually had more than one spare in the trunk. Putting an extra piece of rubber over the hole to protect the inner tube or double tubing would throw off the tire balance but that wasn't a big issue on winding Cape roads.

All this rationing could put a crimp in the plans of a teenager who wanted to go on a date unless he happened to know a cranberry farmer. Farmers were privileged to have extra gas and rations and maybe even a worn out tire or two for a favored worker. So clearly there was some advantage to all that hard labor in the bogs. At the time there were fines for solo recreational driving and signs with slogans like "Have you really tried to save gas by getting into a car club?" Do you suppose that slowed the boys down?

A few years ago, my daughter and I visited a local farmer on a third grade field trip from Marstons Mills Elementary School. Wet harvesting of the bog was in progress, but the farmer also had an antique dry harvesting machine, introduced about 1957. I am sure Dad would have much appreciated walking behind a dry harvester rather than crawling on hands and knees through the vines. The "Darlington Cranberry Dry Harvester" patent 2780905 was an interesting contraption with a rotating rake of fingers or combs to snag the berries, a conveyor belt, and a storage bin for the berries. This machine has to tread lightly on the vines so as not to damage them too much.

The farmer cut a berry in half to show us the air chambers that make a berry bounce and float. I kept wondering if the wet harvester that was "combing" the vines would fall off into one of the trenches that surround the bog. The operator was an expert at driving this machine around on top of the flooded bog shaking berries loose from the vine. What a treat to see this operation first hand.

We had a bog next to our house in Marstons Mills and I was amazed to see a tractor backed up belly deep into Mystic Lake with an attached power take-off pump lifting water back up the hill to the bog for harvesting and winter protection. That tractor ran day and night filling the bog with water. As I walked the dog around the bog I also found this to be a haven for ticks. The smallest ticks you have ever seen. I spent an hour picking them off me and the dog.

Once picked, cranberries have to be sorted. This is how they came to be known as bounce-berries. Grandpa 'Kanah was also a mechanic who worked on his own salt-works windmills, so I wouldn't be surprised if he had some experience with a cranberry "bounce board separator." From a patent application: "Hard marketable cranberries bounce off the hard, angled surface, into a marketable cranberry collector. Spoiled or bruised cranberries are soft, impact the hard angled surface, and slide down into a rejected cranberry collector. Repeated bouncing of cranberries on angled surfaces is used to separate marketable cranberries from spoiled cranberries."

There are many more patents for cranberry picking and sorting devices at the U. S. Patent Office. The sheer volume of these patents is an indication of how important this berry is to the world. Maybe you have an inventor in your family?

The recipe for cranberry salad made in our family is often requested by visitors to our home. I remember assembling my grandmother's grinder to turn whole berries into mush—the same grinder that we used to grind up sea clams for chowder. Grind two cups of raw cranberries. Gather ½ medium orange, 1 small can of crushed pineapple, 1 cup of sugar, 1 cup chopped pecans. Dissolve 1 small package of raspberry Jello in 1 cup of hot water and add the other ingredients. Chill until set.

As cranberry salad is served at our table throughout the year we have come to realize that our family has experienced the complete circle of life of the cranberry from planting to picking to packaging to preparing to partaking.

Picking Started Today

by
Donna Truran

Drawing upon family history and traditions, Donna Truran imagines how the entry in the diary of her grandmother, Mabel F. (Besse) Jones, might have read for September 20, the first day of harvest in the year 1908.

Cranberry picking started today. The whole household was up before dawn. Chores had to be done and everyone had to have a hearty breakfast. There was fried cornmeal mush, eggs, applesauce, and coffee. Then I put a pot of soup on so we could have a good dinner when we came in at noon.

Picking the bog is a long and hard job, but we all enjoy it. Days ago Grandpa Besse got the barrels out by the side of the bog. They were all lined up like soldiers. Each barrel had to be filled to the top before he put the barrel head on. He was very careful tallying the measure buckets that each picker would pick. Grandpa hadn't gone to school and he could barely write his name. He preferred to make his X, but he could and did count accurately and would say, "Ought is ought and one is one and by God I can cipher."

That year I spent many hours on the bog. Little Mabel was about three months old and Grandpa could look after her by the edge of the bog. I noticed that she didn't like the bright sun and would fuss and cry, so I made her a little white sunbonnet out of one of Will's handkerchiefs. I used pasteboard slats to stiffen the brim. She was much happier after that.

The young girls and ladies picked with small snaps while the men had larger rakes and scoops of various sizes. Of course, we had to wait for the bog to dry off from the early dew. The berries have to be as dry as possible. It was a good crop that year and the price was $14.00 a barrel.

Grandpa was a finicky old man and he certainly kept his bog well groomed. After harvest the bog would be raked—all in one direction—so that the vines would be easy to pick the next year. That would also remove any broken or weak vines and make more vigorous growth for a better crop.

In the early winter, the planks would be put in the flume and the water from Harlow Brook would fill the bog. Soon it would freeze over solid.

When the ice was safe, the men would go over to the sand pit with wheelbarrows and get load after load of fine sand to spread on the bog. This was an all-winter job. They always left the ice closest to the house without sand until last so that the children could skate.

In the spring, the planks would be removed and the ice that was left would sink down with the weight of the sand. This would leave a nice layer of sand all over the bog. Then the bog would be groomed and weeded, getting ready for the next harvest.

We Always Managed to Have Fun

by
Leonora Fernandes

My aunt had family in Wareham. In 1938, when I was ten, we came down from Rhode Island to stay in her brother's cabin to pick cranberries. My job was to fetch the boxes and keep an eye on her berries to be sure they were tallied properly. Needless to say, our picking was cut short that year because of the '38 hurricane.

When we moved to Wareham my summers were spent picking stones on the bogs for twenty-five cents an hour. Believe me, if you have been on your knees scooping cranberries, you don't want to kneel on a stone. Even if we weren't tormented by rocks, we would sometimes hit a beehive, or a snake—we couldn't get away fast enough!

I set cranberry vines, which after being cut were kept in a reservoir until ready for planting. One hot day they brought the vines to us from the water. They were swarming with small gnats which bit every square inch of flesh they could attack.

When I got old enough I learned to scoop and use a snap on the young vines. We would gather at Cape Cod Cranberries on Carver Road and a truck would carry us to pick all over Massachusetts. You had to deal with a sore back and knees and hot sun. The only reprieve was if it rained.

Sometimes if they flooded the bogs the night before because of frost we had to sit and wait for the vines to dry before we could begin picking. The teeth on the scoops were made of wood. Moisture would make it hard to pull the teeth through the tangled vines.

We loved the mornings waiting for the bogs to dry; it gave us a chance to play baseball on shore.

I graduated to be a tally keeper. I had a list of the pickers' names and had to keep track of how many boxes each one picked and check each box so that the wheeler could carry it to shore. Sometimes at the end of the day there would be an argument if one person had a higher count; they usually settled with a competition the next day.

Weeding was another job I did. We had to walk carefully so as not to damage the blossoms. And we had to watch out for poison ivy.

Working on the bogs was only one of my jobs. In June we would stay in Falmouth and pick strawberries, then move to Maine in July. Their berries ripen later. When we arrived home there were always blueberries and raspberries.

My aunt was a widow and she loved and cared for my brother and me. She provided us with a warm, clean, and healthy life. We thought nothing of helping her financially so that we would have proper clothing for school.

We would start school in October every year after cranberry season. Wareham was very understanding. I have no regrets. I believe that when you work for something it makes it more precious. We would still find time to play tag, baseball, hide-and-seek, and jump-rope. There was no television.

I graduated from Wareham High School in 1946 and worked in the Wareham Assessors for twenty years, after my four children were grown up. My husband, Tony, was Municipal Maintenance Director for twenty years. We were married forty-six years before he died in 1994.

A little episode with my aunt's elderly cousin:

I was tally keeper. We had a small section of bog which was in a hollow and not large enough to have all the pickers. I asked the elderly gentleman to wait at the top of the hill while we went down to pick that section. He sat on a cranberry box and fell asleep. Soon I heard people yelling; he had rolled down the hill and into a ditch.

I hid my face behind my tally board and couldn't stop laughing. When I gained my composure I said to him, "Did you get wet? Didn't I tell you to stay up there!"

1935: A Photo Essay

by
John B. Howes

The Secret Passage

by
Cynthia Haskell

There were five bogs behind our house at 16 Randall Road, Mattapoisett, off Route 6 near Turk's Restaurant. They were owned by George Eldridge, and later purchased by Edwin Parks of Middleborough. There was a series of two bogs north of our house, and what we always referred to as the "secret passage:" an irrigation ditch that provided gravity-fed water to the two bogs behind our house, and by underground culvert to a bog south of Route 6.

We spent many a winter skating the secret passage to the north bogs and playing among the saplings that were frozen in this passageway. A very dear friend of ours whom we nicknamed Uncle Bud was a wonderful skater, and would often skate with us after work. We would all hold hands as he skated really fast and whipped us around, all of us girls screaming with joy. The smaller kids he'd skate backwards and help until they learned to stand on their own. He was a wonderful teacher, with a lot of patience, and would explain to us why we needed to buy skate guards instead of walking up the long driveway to our house on our skates, to keep our blades sharp.

We often dreamt of the day one of us would become a champion figure skater!

In late fall the Bowman family from down Aucoot Road would come by horse and wagon along Route 6 to our road to pick cranberries; the bed of the wagon would be filled with children. We would run down the driveway behind them, begging to jump on for a quick ride to the bog.

In the fall after harvest, when the bogs were flooded, my next-door neighbor and I (being a tomboy) built a raft with two drums under an old sheet of plywood. We pretended to be Robinson Crusoe exploring the ocean as we poled our way across the bog, teetering all the way. I was usually the one who ended up in the water, not he.

We had a lot of fun on those bogs. We respected the owner, and never picked a cranberry without asking first, and he would often let us pick a handful or two to eat or pick a gunny sack to take home to be made into cranberry nut bread or relish.

Life and Death at a Cranberry Bog

by
Michael Houghton

In the late fifties, our family lived near a cranberry bog in West Yarmouth. Getting to the bog involved a fairly short walk through the woods behind our house. The woods and the cranberry bog were a magical playground for us. In the woods we built a fort with an underground entrance, played games, climbed trees, and even made a trail so that we could race our bikes. The bog opened up even more opportunities for fun and games.

One of our favorite games included sneaking up to peek at the migrant workers, whose job it was to harvest the berries. It was common to see them ferried around, packed in the back of a truck whenever they left the bog. Occasionally we would run into them at the Great Island pharmacy or other stores in the small shopping center close to the bog.

Interestingly, we were led to believe that they were Portuguese, because that was the language they spoke; but years later as an adult I discovered that they were actually from the former Portuguese colony of Cape Verde.

The housing consisted of two large ramshackle shacks, practically devoid of furniture and complete with outhouses. In the summer and fall the shacks would be filled to the brim with the all-male migrants. It was a stark contrast to the post card image of Cape Cod, with its beautiful beaches, motels, miniature golf, gift shops and restaurants. I can't say for sure, but my guess is that there was no running water in the houses.

Years later I was driving through Appalachia to get to Matewan, West Virginia, where I was shooting a corporate film about a flood that happened there. As I passed through the mountainside, the view was dotted with dilapidated shacks, outhouses, and people living in abject poverty. I saw people standing outside their shacks cooking meals in pots on an open fire.

It brought back vivid memories of the workers at the cranberry bog. I considered the irony of all those summer tourists driving in bumper-to-bumper traffic along Route 28 in West Yarmouth looking at the scenic sights and heading to the beach. Yet less than a half mile away down a dirt road there were these workers living in such harsh conditions.

The caretaker of the bog was a Cape Verdean named John. He would stay there year round and lived a rather solitary life in the winter. One of his responsibilities was to prevent us pesky kids from inadvertently damaging the cranberries and playing with the heavy metal rail cars that the workers would fill up with the harvested cranberries. The carts would be coupled together, filled by the workers, and would roll along a small rail track to where the cranberries were then unloaded. After the season was over and the workers had departed, the train of carts remained motionless on the tracks until the next harvest.

Well, not exactly motionless. We figured out how to couple the carts into a train. We would push them to get them rolling then jump aboard, mimicking the hobos we saw in the movies who would get around by jumping on and off trains. The risk of injury was high, but as kids we didn't care about that. It was about having fun and acting out our childhood fantasies, with real-life props.

Our fun seemed to end the same way each time. John, the caretaker, would return to the bog and chase us away. He was an older man—at least he looked old to us young kids. But I guess working on the bogs makes you strong, because when he chased us he would sometimes nearly catch us.

Another favorite activity was to pretend that we were soldiers doing recon on an enemy base camp. We would sneak up and see who could get closest to the bog without being spotted. I remember one time one of the workers spotted us as we crawled along the ground and hid behind trees firing our pretend stick rifles. He apparently was amused at the sight and took a giant pork chop that he was holding, positioned it like a machine gun and pretended to fire back. That summer we immortalized him with the nickname "Pork Chop."

Another time I was sneaking up and had reached the side of the outhouse when suddenly the door opened and an African-American woman with a pink wig, tight-fitting dress, and high heels came out. I stood frozen because I was standing very close behind her. I even held my breath so that she wouldn't know I was there. She strolled back to the shack without any glance backward, but I stood frozen and in wonderment because that was the first woman I had seen there, and even more curious to me was the fact that she had pink hair.

Now that I think back to that incident, maybe the workers weren't living as much of a solitary life as I thought.

In winter it was common to walk through the woods to skate on the frozen bog, which the owners had flooded to protect the vines from winter kill. It was the one time of year when the bogs seemed to be open to the neighbors. No one ever appeared to tell us to leave or chase us. That was a good thing, because running through the woods while wearing skates is probably a skill that none of us could have mastered.

We would skate on the bog, play hockey, and sometimes even swim when on occasion someone fell through the ice.

One weekend after a fresh snowfall two of my friends came knocking at our door. "We just found a dead guy in the woods!" they excitedly told us. My sisters and I thought it was either a very bad joke or possibly an exciting, potential adventure. So off we went through the snow and they led us to the shacks at the cranberry bog.

That trek through the woods to find a dead body turned out to be a life-changing moment: the loss of my childhood innocence. Because, as we turned the corner of the shack, there was John the caretaker, lying dead under the eaves of the house, his face purple, eyes open with icicles dripping down. At that moment my sisters, friends, and I had come face-to-face with the grim reality of death. To this day I can close my eyes and vividly see the gruesome sight.

He was covered in blood, and there were footprints in the snow leading down the road. Our imagination ran wild. We instantly surmised that he had been murdered and the killer was lurking nearby. So we ran back through the woods and told our parents, who then called the police. Later we learned that poor John hadn't been murdered, but had hemorrhaged internally. He had attempted to walk down the snow-covered road for help but had to turn back, and collapsed where we found him.

That day, the cranberry bog ceased to be a source of fantasy and fun. As I look back, I don't think any of us talked much about the incident for years.

Dry Picking

by

Joyce L. Morrison

In 1967, when my boys were ages four and five, I was offered a job that was ideal for a mother. You worked only when it was dry and sunny, and you could bring the children with you. Dry picking of cranberries is done in the fall, the best part of the year, on the nicest days. What more could I ask?

We were mostly a female group and some of us had children. We would get a call to come to the bogs around 10 a.m., after the dew had burned off and the bogs were dry. The children would play along the edges in the sand hills. We would bring picnic lunches. Some of the children went to kindergarten and were transported in turn by the mothers.

The pay was very good: $2.10 an hour, and if we stayed the season we received another fifteen cents for time worked. I always stayed the season even into late fall sanding.

Dry-picked berries are sold as fresh fruit. The picking machine is like a large lawn mower with very long teeth. It moves along by itself with an operator guiding it. The machine combs the vines, and the berries fall into the box. The bog is picked in a circular pattern following the vine growth. When the box is full, the operator slips it to the ground. Before picking started each day, the box boy would have placed empty boxes here and there over the bog. At the end of picking all full boxes were picked up and taken inside to stay dry.

Dry picking was very enjoyable work for me—being outside, communing with nature. The sun baking down on your head. The snakes combed into your berry box. SCREAM. The biting bugs. The mouse family who squeaked loudly on every pass over their nest.

Every job has screams and squeaks, but you learn to cope. I have done lots of kinds of work over the years, but those golden days on the bogs were the best.

Cranberry Pond

by
Faye George

North Weymouth may have had ocean front, but the rest of us landlocked citizens could claim our ponds. In South Weymouth there was the reservoir, Great Pond, where a pre-Columbian dugout canoe turned up in the 1960s drought; in East Weymouth it was Whitman's Pond, where I first learned to skate; and Whortleberry, which was located halfway between them. But heading northwest, above the Landing, you'd find Cranberry Pond.

We had been casually house hunting for about a year when the hip-roof colonial on Broad Street came on the market, a house that, in euphemistic real estate speak, would need "some updating;" nothing my carpenter husband couldn't handle. Anyway, what mattered most was—location—location! And I knew that behind that house and through those woods lay Cranberry Pond winking in the sun and waiting for us. Cranberry was one of those glacial kettle ponds that the ice sheet dug in its slow slide south. It is highly unlikely any serious cultivation of the wild cranberries that gave the pond its name was ever attempted. The surrounding terrain was not nearly as conducive to commercial cultivation as the expansive marshy acreage of Plymouth County towns, father southeast.

My first priority for raising my family had been—woods and a pond; the house itself an afterthought. Though it may sound counter intuitive, wilderness is civilizing; I believe the best way to raise children to become responsible, humane, sensitive human beings, is to introduce them early to their first mother—nature. Give then love, a healthful diet, good books, woods and ponds—an intimate acquaintance with the natural world—and they won't go wrong.

Cranberry Pond held all the lure a child would need for endless entertainment and exploration. There was always something new to discover season after season in the fresh outdoor air. Where better to stimulate the young imagination—be an Indian creeping through the brush after game, harvesting wild cranberries for pemmican; no matter that vines were spare or the fruit the birds left basically sub-edible. The pond, though not much for fishing—sunfish and minnows, a few pickerel—was nevertheless an open-air zoo: crayfish, frogs, snakes, turtles, salamanders: the myriad varieties of amphibious life, with dragonflies and water striders darting over and around water hyacinth, pickerel weed and pond lilies. Here the children would learn how pollywogs grow into frogs, where turtles bury their eggs, that the safest way to save a snapper from becoming roadkill is not what you'd think, with a stick; you pick it up by the tail. The place was a haven for birds as well, many of the varieties fast disappearing in the rapacious development of open space: brown thrasher, flicker, grosbeak, indigo bunting, oriole, redwing blackbird, scarlet tanager, towhee, waxwing, wood thrush. . . lured by the bounty surrounding the pond; and here we would get to know them.

The pond was not for swimming, but it came into its own with the first good ice of winter. It was shallow enough to freeze solid after a decent cold spell, which made it safe for children. I'd have the makings of supper prepared and in the refrigerator by the time the children came in from school, and grabbing our skates from the box in the mudroom, we'd be off, with Sunday our beloved cockerpoodle at our heels. We'd cut through the break in the stone wall that marked the end of our yard from the woods, going on through to the right-of-way under the Boston Edison power lines and down the footpath flanked by dry stalks of Queen-Anne's-lace, milkweed, yarrow, and goldenrod, quickening our steps as we approached the clamor coming from the pond. Shrill voices in the crisp air, the smack of hockey stick against stick against puck, squeals and screeches, giggles, guffaws, the thuds and groans of the hapless—all of it a song sung in praise of winter.

Finding a suitable spot to sit, we'd put on our skates, beginners needing some help from Mom with lacing-up, then after stashing our boots in a safe place, sweep out onto the ice. The alder bushes poking out of the ice at shore-end gave the novice skater something to grab onto while mastering the art of balance and controlled momentum; also providing Sunday dog with endless opportunities for olfactory exploration despite the challenge of maintaining canine dignity with four fluffy, uncontrollable paws distinctly unaccustomed to ice.

How much better out here in the open air than enclosed in the steel shell of an indoor rink, dank extension of the sterile, artificial world enticing us away from the real thing, an honest-to-goodness outdoor pond in the middle of organic, vital nature. Nature—that alone can give the visceral knowledge of what it means to be human, and where our place is in the intricate chain of being.

On shore someone would eventually get a fire going in the ring of blackened stones, where skaters took turns warming hands; sometimes a bag of marshmallows would magically appear. One could imagine a Shawmut hunting party huddling there crouched over the fire, roasting a pheasant or wild duck on a stick, a few others perhaps over combing the vines for the last of the cranberries…

I would not need a watch to tell me when it was time to get on home and start supper. By then the children's ankles would have had enough, their toes cold, fingers numb under their ice-beaded mittens, the last of the Kleenex used up, and appetites well whetted. We'd shuck our skates, plunge cold toes into colder boots, and follow Sunday dog's lead back down the path with the sky behind us bleeding a gorgeous cranberry red through the black calligraphy of trees.

The Model A Bog Buggy

by
Kirby Gilmore

Henry Ford produced his Model A from 1928 to 1931, and this vehicle was no doubt the toughest and most durable automobile ever built; yes, you VW Bug fans, the Model A will be around when your Bug has long become a pile of rust!

Following the war in 1945, cranberry growers were buying up all the Model As they could find to convert to a new life on the cranberry bogs. Joe Chamberlain even went as far as Maine in search of a few good vehicles. You see, by this time the Model A was considered to be old-fashioned, and everyone had a hankering for the new models now that the war was over. Model As could be bought for as little as $15.00. Every grower had to have one, or several, to run chores about the farm, to gather in the harvest, or for sanding during the winter.

To make a buggy all you had to do was bring the Model A back to the screenhouse, chop the body off, and replace it with a wooden homemade bed that could tip like a dump truck. Hundreds of Model As were cut up and placed into service, much of it with crazy young men at the wheel hot-rodding about the bogs much like they did with Jeeps in the war.

This Model A buggy was owned by H. R. Bailey of South Carver; he owned bogs but also manufactured the famous Bailey Cranberry Separator, as well as wheelbarrows, dusters, pumps, etc.; many of which are in use today. This "buggy" sat at his shop for fifty years, untouched as it was from the 1950s.

I replaced the body using period material and fasteners, and added the 6-Volt alternator. It is used daily at Benson's Pond Bog. Some day I may have to replace the original 1930s tires.

Stumps

by
John B. Howes

John B. Howes died in 2004 at the age of 91.

In the 1930's he put together a meticulously researched, 120-page book (replete with photographs, illustrations, and detailed charts) about all the many facets of constructing and maintaining bogs and the growing of cranberries. Bogs: Cranberry Growing in Massachusetts *remains a veritable treasure trove of cranberry culture and lore. Although this extraordinary work was never published, Howes's daughter, Betty Howes Johnson, has carefully preserved the typescript and generously made it available to Rock Village Publishing so that we could include a small portion of it in* Cranberry Memories. *We have selected the following as representative of the contents as a whole. Additional excerpts appear under the chapter heading of* "Wind Machines, Newspapers, and Algal Scum."*—the editors*

Look a prospective swamp over in the fall after the leaves have fallen and while it is bare of snow. Then its shape and size can be seen. The amount of wood on it can be estimated. If it is not frozen, drive your heel into any suspicious looking bumps to

A Swamp Suitable for Cranberries. Situated beside a pond furnishing a good water supply.

determine if they hide large stumps which may be costly to remove. Take a maple or willow shoot and sharpen one end. By pushing this down in various parts of the swamp some idea of the amount of muck underlying it may be determined.

Many times when one is looking for a swamp it pays to take some elderly native who is a hunter or woodsman along. Many times he can answer many questions which come to mind about the water supply and the surrounding land. Perhaps he might even have been one of the men hired to cut the swamp over many years before and so may know something of the stumps to be encountered buried under the surface. If the swamp is land-

A Massachusetts Cranberry Bog

A Swamp as it Appears in Summer.

The same swamp after the leaves have fallen and a good time to examine it.

locked and partially fills with water in the spring, this is an excellent chance to find out how level it is.

Buy a sufficient amount of shore property to insure your water privilege and also to be able to obtain sand in future years. The shore should also be cleared back from the swamp a distance

Building a Maple Swamp

to prevent large trees from shading the bog and to help in air drainage, which often keeps frost from setting in many nights. Many times a large tree near a bog such as a maple may drop so much seed on a bog that the young seedlings become quite costly to keep weeded.

Cutting Brush

Turfing

If the stumps are cleared by hand labor, after the brush is cut and burned the next operation is to turf and remove the stumps as they are encountered. This is the method used in my

Removing Stumps by Hand

A Slow Speed Winch for Stump Pulling

family for three generations. The tools needed are brush scythes, axes, turf axes, turf hooks, grading hoes, and forks. One or two tackles are also needed. By hooking these up together and attaching one end high in a tree, it can be easily pulled over. The stump will come out after most of the roots have been cut.

The roots and stumps are put in piles as the work progresses and burned or hauled off in sleds. Do not forget that

A High Speed Winch for Removing and Piling Stumps

one or two good heavy prys about ten to fifteen feet long will always come in handy prying out stumps and roots.

The machine method varies from the hand method only in the amount of machinery employed. Regardless of the amount of machinery used there is always plenty of hand labor. There are machines designed by bog men specifically for pulling stumps. These vary according to the ingenuity of the builder and his experience with this work. Steam and gas shovels are used to pull stumps as are also heavy tractors and bulldozers. I have seen gas shovels pull stumps with a long boom attachment and pile them on shore all in one operation where the swamp was narrow. The stumps are often pulled ashore and piled with winches and tractors.

Bulldozers, shovels, and tractors and winches with scoops are all used in grading. Trucks have been used in sanding on a swamp which was frozen solid enough to hold them. Gasoline powered trains of cars are often used for sanding. Often the use of any one of the forms of machinery mentioned will save many times what the operation would cost with hand labor alone.

Pulling Stumps with a Winch

Stumps all Pulled

The older people used to remove only those stumps which prevented them from grading a piece of swamp level. This usually meant only the smaller growing trees, while the large pine stumps they encountered were usually cut down a few inches below the grade. This saved the expense of removing them but was still a laborious task as anyone will know who has done this. The roots of these stumps help in drainage and commonly the vines are more thrifty and productive over spots where there are buried stumps. Such bogs must never be allowed to freeze deeply,

Rough Grading

Pulling Stumps Ashore and Piling Them with a Winch

as frost will cause heaving and make the surface rough. This is what usually happens where the stumps are not removed; if there are a great many of them it is usually thrifty to employ machinery to remove them. A stick of dynamite is often put in the

Grading

center of very large stumps to crack them apart so that they can be removed in several pieces.

Maple stumps are most easily removed by cutting the roots as most of them are in the turf and near the top of the ground. If a strain can be put on the stump a very large root can

Opening a Sand Hole

Spreading Sand

Setting Vines

be cut with a single blow of an ax. When stumping a maple swamp I usually cut the trees leaving a stump about one foot high. Then a loop of the cable is thrown around this and a strain put on it. One or two men can then cut a circle around it in a few seconds and it will pull out easily. All the loose turf and mud

The Completed Bog

should then be cut off of it while it is on its side to fill up the hole, before it is dragged to shore.

Big cast steel hooks are used to catch into stumps which have no top, or a root has to be dug out to hook around. Pitch pine stumps have a deep tap root which makes them hard to pull out. Often a maple stump is pulled and an old pine stump is

A New Bog After One Year

found directly under it. The roots of old pine stumps are often locked to one or more stumps and have to be chopped apart with an ax before any can be pulled out.

I have seen many swamps built recently where there were three layers of stumps encountered. First there were maples growing. Underneath these were white pine stumps, one layer cut probably twenty-five years ago and the other cut perhaps fifty or seventy-five or more years ago.

After the stumps are out they may be piled and burned on the swamp or piled up on shore. They can be dragged ashore on sleds or with cables using oxen, horses, winches, tractors or any other ways. When it is possible to burn I think anyone will find it the best to burn the longest hours possible as it usually takes several hours to get a fire hot enough to burn stumps, then the next day this has to be built up again. I usually use a sled to take off the smaller stumps and roots that are easily handled. A track-tractor is the best where the going is soft.

I usually drag off the heavier stumps with a winch and cable. A good strong tall tree is selected on shore where it is desired to put a pile of stumps. This tree is strengthened by the use of guy wires and a block or pulley is securely attached as high as possible on it. The winch is set up behind this tree and the cable run through the block and out on the swamp. The stumps are pulled ashore and piled around the tree. After the pile has dried out somewhat it can be burned.

Feeding Frogs and Fooling Friends

by
Judy Pratt

We used to have a bog in back of our house when we lived on South Meadow Road in Carver. We were the fifth house on the entire road. A pine grove separated the yard from the bog. As children, we had no other kids near us to play with; my mom didn't drive. So we had to make our own fun. Computers and cell phones weren't around then.

My sister and I made little forts or houses—using pine needles for the walls. My dad dug a small hole by the pine grove where water would come up, and we'd have frogs, hence the name, The Frog Pond. We'd put a few plants around it, and when it rained the pond filled up.

We used to go walking by the bogs, and I loved the mayflowers, and little purple flowers with yellow in the middle. I'd watch the hornpout (catfish) in the brook that was used for the bogs. I even fed a frog a cranberry once! That took a little time. I'd put a small, ripe cranberry on the end of a stick and wave it in front of the frog until he took it.

There were box turtles and snapping turtles; you could tell the snappers by their scalloped shell and pointed tail. I liked the turtles with the yellow dots on them. Sometimes we'd see a muskrat swimming when the bog was flooded.

I learned how to skate in the winter on the bog, by pushing someone sitting in a wooden chair. One of my sister's friends went through the ice; he'd got too close to the reservoir. He got out okay—lucky for him our house was close by.

Friends who came from New York once asked how they drained water from the bogs. I told them there was a plug they pulled, and they believed it! I eventually told them the truth.

There were cultivated blueberry bushes growing by the bogs. Sometimes we'd see a pheasant or a deer. One time I saw a poisonous black water moccasin snake by the bushes. Luckily for me it slithered into the water nearby. I enjoyed riding my bike—trying to beat the rain—when I'd hear that warning rumble.

I enjoyed watching them harvest the cranberries, especially the water picking method. The foliage was always pretty around the bogs. I was always respectful of the owners' property, and had their permission to ride my bike or walk around. I always enjoyed spending time with nature; there were so many interesting things!

Bog People

by
Alan Pierce

"But we're bog people*!" my mother informed the motel clerk who would not accept a personal check. She was trying to book a room for my wife and myself who would soon be visiting from Minnesota.*

I grew up in the same neighborhood as the late Ann Liza Morse, author of the *Cranberry Kids* series. Our house on Main Street, in West Wareham, Massachusetts, overlooked the Hathaway Bog, where each summer we caught spotted pond turtles in the ditches, the same ones, over and over; and where each spring the bog offered up a wall of frog sound as mating season revved into high gear. Kids loved to walk the top of the white wooden fence that separated the bog from Main Street.

My father, Bob Pierce, was Bog Foreman for the L. B. Handy Cranberry Company from 1935 to 1952. Handy owned several pieces of bog in Wareham, Marion, and Middleborough.

The "Big Bog" and large screenhouse were located at Blackmore Pond. Handy had two daughters, who became Aunt Dot and Aunt Hazel to me. During picking season the screenhouse was a beehive of activity: my father coordinating activities, my mother running a cellophane machine (heat sealing bags of fresh cranberries), and my grandfather nailing the lids on small gift boxes.

After Handy died, my father went into partnership with Frank Crandon of Acushnet, and the C & P Cranberry Company, Inc. was born. The fifty-acre main bog and screenhouse were located on Charge Pond Road in Wareham. The smaller fifteen-acre S-curve Bog was in West Wareham. A new International pickup with the C & P logo on each door was parked in our yard each night. I was so proud!

Near the screenhouse, a rusted metal vat sat high up on four concrete posts. The previous owner explained that long ago, shade tobacco scrap was shipped from the Connecticut River Valley, placed in the vat and mixed with water. A fire was then built underneath the vat, and the resultant nicotine broth was tapped off and used as an early insecticide.

I spent the next few summers whiling away long lazy days at the main bog with friends Bobby Stevens and Eddy Grassi. While my dad worked, we would explore the local environs, pole rafts in the shallow reservoirs, or hike up to Charge Pond and spend the days swimming and diving in that magical place.

On several spring and fall evenings when frost threatened, I accompanied my father to the State Bog (the University of Massachusetts Cranberry Experiment Station) in East Wareham. There the growers sat in a crescent of wooden chairs, wrapped in a cloud of tobacco smoke. Old Doc Franklin sat in back near the large wood stove, while George Rounseville hunched at the front desk making calculations based on information he received from the Weather Bureau in Boston.

Robert A. "Bob" Pierce, 1947

Eventually, George would swivel his chair towards the growers to deliver his verdict, and a great hush would fall over the crowd. "No frost," and the growers would amble out to their vehicles and head for home. "Frost," and the growers would hurry to their vehicles and head for their respective reservoirs. At the reservoirs, they would "pull hard," yanking planks from the flume with a plank hook, which sent water cascading down the main ditch.

This was before permanent sprinkler irrigation systems were installed, and the bogs had to be flooded for frost protection. Often one had to gamble, to flow, or not to flow, as the reservoirs held a limited amount of water, and the bogs had to be drained the next day to prevent damage to the blossoms or berries from oxygen depletion. The C & P main bog deep reservoir was good for three flows.

In my early teens I joined the picking gang each fall, and earned fifty cents a box scooping berries for C & P. My father supervised the harvest activities, and my mother was employed as tallykeeper, marking filled boxes with chalk and noting who had picked them. When our scoops were full, we would dump them into open quarter barrel boxes. It took three of those quarter barrel boxes to make one barrel, as the berries had to be heaped to count as a full box, and three heaps equaled one box. Unfortunately, the onset of picking time coincided with the beginning of school, so I could only pick after school and on Saturdays.

After picking season, the bogs were flooded to protect the vines from winter kill. When the water froze, we took to the ice to play hockey. Bogs were relatively safe places to skate because if you "fell in," you would usually end up with wet ankles, unless you "fell in" over a ditch. In West Wareham, the boys would check the ice condition of bogs near their homes, then compare notes and decide where to play. I was responsible for reporting on Bart's Bog, about a quarter mile from my house, as the Hathaway Bog was among the few which were not flooded for the winter.

When I was fourteen, my father told me, "You can consider this your last summer as a child." So, for the next several

Alan Pierce being taught to drive by his dad.

summers I worked at a number of cranberry-related jobs. I ran pumps and moved irrigation pipe as C & P battled a severe drought. I gave tours at the Ocean Spray Plant in Onset, wearing the mandatory cranberry red shirt and white pants. I toiled on the bogs six days a week running a weed clipper and mucking out ditches with a potato digger. A sun worshipper, I worked shirtless and hatless. On Sundays I went to the beach and drew stares at my bronze torso and fish-belly white legs.

"All The Cranberry Juice You Can Drink For 10 Cents" proclaimed the sign at the Cranberry Welcome Shop in Onset, and despite our supervisor's orders to the contrary, we "juice jerks" encouraged our customers to try for the record. Near summer's end, the record stood at twenty-something cups of juice when two of my high school classmates showed up. Ken tossed down thirty-three cups, but Jay tossed down thirty-four and became reigning champion. He then fled to the patio and deposited his winning volume into the shrubbery. Ken found this extremely amusing, but a few days later he quit his summer job early, as he was suffering the effects of not making a deposit into the shrubbery.

In college I worked summers at the State Bog as a research assistant, investigating the impact of plant parasitic nematodes on cranberry yield with my mentor, Dr. Bert Zuckerman. The station hummed along under the watchful eye of its director, Dr. Chet Cross.

Finally, in 1969, I returned from the Midwest to work my first and only complete picking season, running a Darlington picker on C & P's bogs. After the 1973 harvest was in, the bogs were sold, the C & P Cranberry Company, Inc. was dissolved, and I was no longer Bog People, or am I still?

Living His Dream...

by

Dolores (Sylvia) Paterna

A desire to own a piece of land prompted Frank Sylvia's entrance into the cranberry business. A young man when he arrived from Portugal via "the boat," Frank worked at a variety of jobs, including construction, but farming quickly became his love.

While working for someone else on a bog, he concluded that the only way to get ahead in life was to work for yourself. He went to see a man about buying his own piece of land to farm. That man said, "Why don't you buy that four-acre bog that's for sale on Route 140?" When Frank told some others with whom he worked about this idea, they laughed and said he was just a "small, little man" who didn't come from this country and couldn't speak "good English." His only reply: "I want to learn."

Fortunately, he saved the money to buy that bog, living in a run-down "cottage" during the summer so he could get a feel for the land. During this time, he met and married the former Sylvia Cashiero, who would share in his dream to do well in America. They had twin girls and, eight years later, another daughter. To celebrate these births, he added onto the bog from the money he and his wife had saved.

In the late 1950's, Frank bought his second bog, about three miles away, from a retired cranberry grower. Now, it was a full-fledged operation with his wife in charge of all hand pickers and his sister-in-law, Anna, in charge of the scooped boxes. He later met up with his doubters, who now wondered how Frank had done all this. His simple answer: "If you want anything in your life, you save your money, and you live without."

The bogs continued to flourish with Frank leading the way until his untimely death in 1996. At age 95, Frank had more than fulfilled his dream, and he died in the only proper way—while driving his red Dodge pickup truck around his beloved bog. Now, his children strive to continue his legacy and to do tribute to a man who loved the soil and treasured what life in America had afforded to him.

The Bold, Sweet Taste of Home

by
Aida Elder

Growing up in an immigrant Italian family meant that some part of every day was dedicated to cooking. The fruit I was prepping on this day was something my southern Italian relatives had never seen, and at an early age I had come to appreciate its unique flavor. I knew that the good berries would float to the top when submerged in water for a thorough rinse. My thumb and index finger recognized the firm sensation of a well-ripened specimen when steady pressure was applied—too much give revealed hidden decay. A gentle rub with my thumb over the end would remove any resilient stems.

The clean, red jewels that remained after my vigorous inspection were then split down the middle with a sharp paring knife. The crisp, white center was adorned with a star pattern of tiny golden seeds, contrasting beautifully with the ruby skin. I prepared this fruit with such great care because it was the key ingredient to my very favorite pie of all.

After settling in southeastern Massachusetts, my parents learned to incorporate cranberries into our usual Mediterranean fare. Creations included cranberry pie, cranberry muffins, cranberry bread, cranberry sauce, cranberry coffee cake...and the list goes on. But this fruit enriched more than just our diet—it flavored our New England lifestyle.

Cranberries were a dominant industry in our area, and bogs were a requisite part of the landscape. The very seasons could be defined by the appearance of the bogs. Earthy red tones heralded the beginning of autumn and the harvesting of these petite berries. Flooded, frozen bogs were a child's winter wonderland; to this day the sight of a frozen bog elicits in me an incredible urge to strap on a pair of ice skates. I spent countless hours teetering on double-bladed skates, chasing my older siblings on these make-shift rinks. As temperatures warmed and the frost-repelling waters were drained, the vines of a bog in spring flaunted delicate blossoms that burst into life. The fertile vines of a summer bog were a playground for deer, birds, frogs, turtles, and a host of insects.

I took for granted how much cranberries had shaped my life until I left home to attend college in the city. Cranberries were as ordinary as, say, quahogs, but the confused look I saw on people's faces when I mentioned my summer job at the UMass Cranberry Experiment Station were strikingly similar to the reaction I received when uttering the name of my favorite hard-shelled clam.

"There is a *cranberry* experiment station?"

It was as if such a small, unassuming fruit did not deserve this distinction.

My summer job at the Cranberry Station was the most fun I have ever had being employed. Established as a state-funded institution to support the Massachusetts cranberry industry, the Cranberry Station staff conducts research and provides outreach to local growers. Working for the plant pathology lab involved outdoor assignments practically every day—collecting specimens of fruit rot, organizing experiments with plots of variables and controls, assisting local growers, and sometimes just plain getting dirty pulling weeds and resurrecting an abandoned bog. Not only did I learn important scientific principles, I also grew to appreciate the kindness and good nature of the station staff, as well as the local cranberry growing community, and by the end of the summer, my tan was the envy of all my college roommates.

Hours spent at the Cranberry Station made me understand the hard work required to cultivate a successful crop of cranberries. Between weeds, fungi, and invasive insects, it's a wonder berries grow to maturity at all. The sight of a bog that is showered with sweat and love is immediately evident, for within the crisp borders is a vision of green, without patches of weeds or fungus to disrupt the uniform color palate. After working at the Cranberry Station, one could not help but develop maternal feelings toward these rectangular collections of vines. My sister, who also worked summers at the station, while driving past a bog that had succumbed to weeds and fungus twenty years later, instinctively sucked in her breath and exclaimed, "Oh, how sad—it just needs a little love!"

I recently moved back to southeastern Massachusetts after many years away. I asked my mom for the recipe of that pie I remembered so well, and for which in all my travels I had found no equivalent. It was with great excitement that I prepared the berries and followed the recipe, wondering if the final product would match the beautiful flavor in my memory.

The aroma wafting from my kitchen signaled the pie was ready. I peered into the oven, admiring the candy red juices that had seeped from the crust. Unable to wait for very long, I cut into the flaky shell and served myself a bright, oozing wedge of cranberry pie. The bold smack of the first bite surprised me; this unexpected, complex burst of taste compelled me to take another. The beautiful marriage of tart and sweet was a harmonious chorus of antonyms, representing New England in a pastry form—harsh winters and warm fires, hot summer rays and cool ocean sprays, strong characters and soft hearts. My taste buds quivered with delight.

I had indeed arrived home.

Cranberry Pie

1½ cups sugar
¼ cup flour
¼ tsp salt
3 cups cranberries, cut in half, or very coarsely chopped in a food processor
½ cup water
½ tsp orange extract
pastry for a two-crust pie

Preheat oven to 400 degrees.
Prepare crust and line an 8-inch pie dish with dough.
Combine the sugar, flour, salt, cranberries and water in a 4-quart sauce pan. Bring mixture to a boil and cook for 5 minutes, stirring constantly.
Remove from heat and stir in orange extract.
Pour cranberry mixture into prepared pie dish and top with remaining dough. Cut slits into top for venting of steam. Brush dough with milk or egg whites.
Bake for 35 minutes, or until crust is golden. Serve slightly warm.
Makes one 8-inch pie.

Where was I—Mars?

by
Wayne C. Christiansen

Haling from the vast farmlands of Iowa, my eyes were accustomed to seeing, from horizon to horizon, endless rows of corn, soybeans, oats, and hay. Roads dividing the fields were aligned in grids that ran due north, south, east, and west. Farm work was done with equipment resembling that used to construct super highways. Crops started out green and wound up shades of brown.

I grew up thinking that's the way things are.

Then I married a New England girl, settled here, and saw my first cranberry bog. It was autumn and we were on our way to Plymouth with our two daughters, when abruptly we broke free of the trees. There on both sides of the road floated a beautiful carpet, like crimson foam. My wife identified the sight as a cranberry bog flooded for harvest.

My mind reeled. What? Things are grown in a bog? Impossible! A bog was something you had to drain to make room for more fields. A bog meant that there was too much water for a crop to survive. I had seen rice paddies up close in Vietnam in 1969, so I understood the concept, but never expected to see anything like it in Massachusetts.

And the color! What kind of crop is, well…red? Where was I—Mars?

I stopped the car and we walked to the edge of the bog. I poked at the red balls and watched them bob up and down in the water like tiny apples. Somehow it didn't fit with what I knew cranberries to be: a bitter, jellied mass pushed out of a can, sliced, and slid onto a plate for Thanksgiving dinner—left to be ignored by all but the most daring, or the most naive. It was a "dare ya" moment when my brother and I challenged each other to eat the cranberry jelly, followed by a quick swallow of milk. The sour, bitter taste contorted our faces in grimaces that never failed to make us laugh. Getting a younger, uninitiated cousin to try was even more fun.

I stood and wondered how this colorful, bright little berry could become the dark red gelatinous glob that loitered in our refrigerator for days before being finally, mercifully, thrown out.

I know what you're thinking now: Cranberries have many uses that Midwesterners could never imagine.

True. And believe me, I have gamely sampled most of them. I've strung them up for decorations, drunk the juice, tasted the sauce, sipped the mixed drink called a "Cape Codder," and well…sorry. Can't stand any of it! Cranberries just aren't for me, I decided: a wonder to the eye, but not to the taste buds.

Years later, some of my family paid a visit from the Iowa prairies. We took them to see Plimoth Plantation, and of course the Rock. I was able to relive my own experiences as they oohed and aahed over the red rectangular bogs. It was on that trip that, ever hopeful, I once again tried a cranberry produce—chocolate-covered cranberries. My response could be summed up in one word: yum! At long last, I was able to redeem myself to the little red berries of my adopted home.

The Chopping Brigade

by
Kathy Becker

When I was a little girl growing up in the 1950's in the Midwest I got cranberry juice only at Thanksgiving and Christmas Eve, which we celebrated at my Aunt Rose's. Cranberry sauce was a twice-a-year treat also, and was always the jelly, cylindrical kind. I remember one year when there was some kind of blight that caused a cranberry crop failure and cranberries and sauce could not be found in the grocery stores in Rockford, Illinois.

Christmas Day was spent at my grandparents' farm with grandpa and grandma providing the turkey and dressing and everyone bringing a dish to pass. My Aunt Jo broke with tradition one year and brought a cranberry salad made with ground cranberries, grapes, pecans, raspberry Jello and whipped cream. It became a new favorite.

Years later at my bridal shower my aunts gave me a recipe box filled with family recipes and Aunt Jo contributed her cranberry salad recipe.

The first year I made it for my family's Christmas dinner I didn't have a food grinder or blender and so had to chop a pound of cranberries by hand. You have to freeze the cranberries first to do it this way and it takes quite a long time. My husband really liked the salad so the next year I enlisted him to help with the chopping.

When my children got old enough to wield a knife they joined the chopping brigade. Then one year my son Nathaniel gave me a mysterious Christmas present which he said I had to open two days before Christmas. It was a food grinder he had seen advertised on late night TV. I guess he was getting tired of chopping. His present still works great.

Aunt Jo's Cranberry Salad

Grind 1 lb cranberries and add 2 cups sugar.
Refrigerate overnight.
Dissolve 1 small package of raspberry Jello in 1 cup hot water. Set aside.
When Jello is syrupy add: cranberry mixture, 2 cups of red grapes cut in halves, 1/3 cup chopped pecans.
Whip ½ pint whipping cream and fold in.
Refrigerate overnight.

Cranberry Connection

by
Amelia Amado

As far back as I can remember I had a cranberry connection. At the age of sixteen I began working on the bogs side by side with my mother. At this time cranberries were hand-picked using a cranberry scoop or a snap. It was very labor intensive and we worked eight to ten hours a day, paid by the number of boxes we could fill.

Along with picking the cranberries we also planted vines. Trucks would come by and pick up all the workers, fitting twelve to fifteen people at one time. The truck had wooden seats in the back. Singing on the way to work and on the bogs was a favorite pastime. Often we were transported from bog to bog to work. Long-lasting friendships were made and romances blossomed.

After awhile my father purchased several bogs in the Harwich and Tihonet area. All but the Tihonet property was eventually sold off. The Tihonet property today is non bog land but holds many memories.

As I had children of my own they carried on the cranberry connection at a different level. After the picking season and entering winter they could not wait for the bogs to be flooded and the skating and bonfires to begin. During the spring and summer bogs were also a place of fun for the local kids. Ditches were jumped and pollywogs were captured as well as other creatures that called the bogs home.

I am an avid collector of cranberry memorabilia; my collection contains my father's cranberry snap and several cranberry labels from various periods. I am now eighty-four years old and my love for the cranberry connection continues.

Onset Memories

by
Mary Vieira Rose
as told to her daughter June

Even now, each fall my thoughts return to the cranberry bogs and how our people looked forward to this season. Neighborhood men and women were signed up by the growers early on; the growers supplied the scoops, which were used during the day and stored in the screenhouse after work.

In August, some growers had workers pick the still unripe berries, green and white ones, called vinegar berries as they were used in the production of vinegar. In mid-September, when the berries were ripe for scooping, the growers sent out trucks to collect the mostly Cape Verdean pickers in the community and bring them to the bogs.

If the bog was wet, we would all sit on the side and wait for the sun to dry out the vines. The workers would tell stories to each other, smoke their pipes (even women smoked pipes) and gossip about local goings-on. Then the "okay" would come from the boss. The workers were quiet in testing their scoops by slowly gliding them into the vines with a rocking motion; they would hope that the vines were good to work on.

They moved forward on their knees in a pattern of scooping in front of them, in and out, then turning from side to side, scooping all the while and reaching as far as possible. If the vines were good, a picker had a good day, by just working in the above fashion. But some bogs had long vines, and the picker had to be extremely careful because these long vines could entangle the teeth of the scoop and break them off. A broken scoop was useless until repaired.

There was no segregation in the cranberry picking business. Everyone worked together; men, women, and children worked side by side happily, all looking out for one another. Sometimes the filled scoops were heavy for the women, so the workers would bring in by wheelbarrow the boxes to be filled, so that the women would not have to carry the scoop.

It was important for the pickers to mark the boxes they were filling, because payment was based on how many boxes they picked. No vines were to be in these boxes, so the pickers had to check their scoops and boxes throughout the day. They would kneel down to start scooping the cranberries, check for vines, carry the filled scoop to the box, empty the berries into the box, check again for vines, and return to their section of the bog, and start scooping again.

When the box was full, the picker notified the tally keeper, a woman sitting on the bog with a pad and pencil. Her job was to keep track of how many boxes each picker scooped; her tally determined each picker's wages for that day.

I remember picking cranberries as a child because of the depression; there were no jobs, everyone was hungry, families were large, and all healthy members tried to do something for sustenance.

My father had his own grocery business, a large country store run out of a free-standing building in the house that I live in today. Above the building his name was printed in huge letters. I learned to read his name and I was proud. I was about six years of age when I used to go with my dad to the bogs for delivery of produce by horse and wagon.

My father brought food to the White Island bogs, which at that time were completely surrounded by trees. The growers had workers who actually lived on the bog land year round, Cape Verdean men who tended the bogs seasonally. My father brought lots of food, tobacco for their pipes, rice, beans, etc. He went on Saturday, their payday, and the men would give me one dollar for helping unload the wagon. A dollar was big money in those days. The men purchased enough to last out the winter, as they lived in those small houses cooking, playing cards, and keeping the bogs in good shape for the growers.

On clear days, they would travel on foot to Rochester and other local areas to tell people news of the community. In today's terms, they were mobile news reporters.

The economic crash of 1929 was felt by everyone, especially business owners. My father's grocery business suffered, as many neighbors could not pay for the produce he was letting them have on credit. So our store closed with many others. For nineteen years my father ran his store, and now he found himself preparing to go to work on the bogs—the only job left for many.

The CIO Union came to Onset, and the officials set up at a building called American House. A man named Frank Santos owned the building. He also had a grocery store on the left side, facing Onset Avenue. The right side also faced Onset Avenue, and people entered through here. The CIO held meetings during the month of September. The growers felt that the pickers wanted more money and that a strike was imminent. This situation had to be resolved quickly, because the berries would rot if not picked, and no one would make money.

The growers held off; there was trouble. The rumor was that there was police involvement, and a neighbor's son lost two or three fingers in a fight with the police. Details are scarce. As all of September went by, the growers lost the September crop, but in early October they sent the recognizable trucks to pick up pickers in Onset, despite the growing strike by workers present on the bogs. Pickers who got on the truck were stoned and called scabs. The police escorted workers to the bogs. Those Smith-Hammond bogs are still visible today in back of Benny's on Cranberry Highway.

The Town of Wareham gave permission for children to pick cranberries. I begged my father to let me go with him. I remember the police escort, and being afraid. One of the bog owner's sons was in Wareham High School with me. We were not in the same class, but he knew I went there. He was handing out scoops. He looked at me and refused to give me a scoop, thinking I was underage. I told him repeatedly that I was old enough, and I finally got a scoop.

I felt sorry for my father because it was so difficult for him to scoop. Everybody else was in the natural position: on their knees, gliding the scoops in front of them with a rocking rhythm moving right along. My father had to stand due to a frozen right arthritic knee and had to bend over to scoop. He scooped in his own rhythm and was determined to fare well. He taught me how to scoop, and again I felt proud, to be able to scoop rather than to have to run and pick leftover berries behind other scoopers.

The owner frequently visited the bogs late in the day while picking was still going on. He would inspect areas already picked by spreading the vines with his feet. If he saw any berries still secretly snuggled in the vines, he would whistle to whoever appeared responsible for that row to return and repick the area. If you argued or refused, you were instantly fired. No cranberry was intentionally left to rot. They were too precious.

Men carried the filled boxes by wheelbarrow to shore, where they were loaded by hand onto trucks and driven to the screenhouse. I see many old screenhouses still standing near the bogs, falling down through neglect.

When I was about twelve years old, we picked cranberries by hand on the weekend. On certain occasions, special bogs were picked by hand, mainly by mothers and children; men could come if they wanted to. This was hard work. Very early on Saturday and Sunday the truck would visit the neighborhood searching for pickers and bring us to the bog, where with very serious attitudes we would wait for the vines to dry. There was no horsing around, because we understood what money meant and were glad to earn some for our family.

The women usually wore huge handmade aprons, which they filled by picking the cranberries by hand. When the apron was filled, they would twist the end and hold it up like cradling a baby and empty it into the box. The bogs to be hand-picked bore fatter cranberries, which the owner did not want bruised by the scoop.

The first day, our hands felt good. But by day's end, the back of our hands were badly scratched. Our fingernails bled, and if we worked the next weekend, the bleeding would begin again. There were no Band-Aids then. The old-timers would tell someone to scrape pine trees for sap and help us put the sap under our nails to ease the bleeding and pain. The young children picked only on weekends with their own little container, approximately five quarts. When we filled it we went to the tally keeper for recording. She sat and watched us all day, and was usually the grower's daughter.

We Onset pickers were transported by the grower's truck to bogs in Carver, also a place called Smalley, all the way up to Norton, Plymouth, Rochester, and Mashpee. In Mashpee, the

Indian boys would stand on the hillside and throw stones at the children picking cranberries, and the growers would have to call the police to make them stop.

I also screened for Ocean Spray in Onset when they first opened. We sat at a conveyor belt where the berries passed and separated the bad from the good. It was usually very cold in the room, to keep the cranberries firm.

I remember distinctly the news one day being heard around Onset: Ocean Spray had dumped a freight car of canned cranberry jelly near the railroad tracks in back of their plant, and anyone could go get them. These discarded cans were all right to eat; they just did not meet specifications. The news arrived here in our village around three-thirty in the afternoon, and our parents let us join the neighbors going to the plant. We took burlap bags and started walking.

We arrived one half hour later to see the most memorable sight ever! We could not believe our eyes. There were hundreds of big cans, just like in the stores today, littered everywhere and piled high. My six brothers and sisters and I each filled our burlap bags as full as possible. It took us a long time dragging those bags, coming over Dummy Bridge; we kept pulling and pulling, being careful not to lose any of the cans.

It was dark when we got home. We were tired, but glad that we contributed to the winter supply of food. There was no work in the winter and everyone waited for spring with great anticipation. Let me tell you, that cranberry jelly went fantastically with jag, and we praised God with every bite!

Work and Play at Edaville

by
Joan Morse (Fava) Fraccalossi

My grandfather, two uncles, and father, in addition to having other jobs, were busy growing cranberries in the 1930's and '40's. My father's bogs were on Doty Street in West Wareham and in South Middleborough. After losing his garage in West Wareham during the depression, my father went to work as a mechanic for the Beaton Company. At that time we lived with my grandfather; my father and uncles would use the separator in his screenhouse to screen their berries.

My uncles would take turns dumping the one-third barrel boxes into the separator. Leaves and vines would blow out onto the floor; these were removed periodically, loaded into a truck, and brought to an area behind the bogs. Pie berries—the small ones that would not be sent to Ocean Spray—would drop into a box. The good berries ended up on a conveyor belt for screening in another room. There were always fifty or more empty boxes in the yard to be brought back to the bogs to be filled again by scooping.

Work went on all year. In spring vines were pruned for new planting, and when the buds appeared there was always flooding for protection against frost. This would take time, because although there was a pump, they depended on gravity for water to flow from bog to bog. My father was often out fixing pumps for other growers when they had problems. My memory of Memorial Day would always be the talk about the black frost and how the men would be up all night.

Sanding wasn't always done by buggies. My father would sand in spring by wheelbarrow. The sand would be shoveled at least three times: dug from a pit, screened onto a Model A pickup, shoveled onto a wheelbarrow and then spread over the vines. Boards were placed on the bog to run the wheelbarrow on.

When I was six we moved to Edaville, where my father was the mechanic. This was an exciting time because the train was just being brought in to use in the cranberry business.

When cranberries were dry-picked, by scoop or machine, the vines had to be dry; it was often eleven o'clock before they could begin. Harvest would last right up until Thanksgiving. The berries that fell to the bottom of the bog when scooped would later be gathered by flooding the bog. These floats, since they were so wet, would be screened in an outdoor separator.

Hoppers, for removing leaves and vines, were located on the second floor at Edaville. On the first floor were the separators, where the berries would bounce on boards, the good ones going on belts to the screening room, where several women would hand screen them.

The screenhouse holds many memories. We played hide and seek on the second floor, hiding among the hundreds of stacked boxes. The first floor was cement, where we would roller skate for hours when harvest season was over.

Our own cranberry bogs were harvested on weekends or evenings after regular hours. It was a family affair: my father, brother, sister, and myself; my mother passed away the first year we moved to Edaville, when I was six. The Model A pickup would be packed with soda, water, and sandwiches, and we were off for a long day of picking. My father's berries were mostly scooped, although when picking machines were invented they were used at times. After the berries were cleaned, they were brought to the Ocean Spray plant in Onset.

There were lots of things to do year round. The men would go hunting, trapping, or fishing. In winter there was ice fishing. Shrubs and trees had to be cleared and firewood cut. Winter was the time for us children to burn brush and cook hot dogs over the coals for lunch! We had fun skating on the frozen bogs and playing hockey. Skating to the music over the loud speaker at Edaville as the train went by was always my favorite time. We spent hours sliding down the many hills and slipping across the bogs.

Summer was always the time to have a boat on the reservoirs, for fishing and paddling about. School projects were often based on things found around the bogs. We gathered wild flowers, and found and identified leaves. My favorite project was collecting bugs and pinning them on a board for science. Wild plants included jack-in-the-pulpit, pipsisawa, partridge berry, and Indian pipe. The worst part was not being able to identify poison dogwood; my sister identified it after chewing on a twig! She was swollen for days. Poison ivy was plentiful but easy to identify.

Spring meant planting new bogs, which was done slowly by hand, using a dibble. Vines were poked into the sand every few inches apart. The vines used were pruned from existing bogs by using knife-rakes. Weeding was always another chore. The only chemical I knew for weeds was kerosene. No one needed a chemical or pesticide license back then. Few of the small farmers had machines to dig ditches; this was done by hand, using a shovel. Flumes were hand-built with wood and cement.

Water was always a worry. When the reservoir went dry due to drought, that would be the end of flowing for frost. Winter kill was always a danger.

These are just a few of my memories. My son continues working my father's bogs in South Middleborough, and I still can't stay away at harvest time.

Extracting Honey

by
Betty Howes Johnson

My dad had honey bees as did most of the bog owners. They placed the hives near the bogs to ensure better pollination.

In the fall when putting the hives to bed for the winter months, the honey-laden frames were removed, leaving some for the bees as food for their winter months. We would bring in the extractor—a metal cylinder about four-foot tall, with a crank handle on the top and spigot on the bottom—into the kitchen and set it on a cranberry box. While Dad and I got jars ready,

Mom would get the kitchen wood stove fired up, warming the kitchen so that the honey would flow more easily.

The frames were placed into the cylinder on the inside near the outer edges. Then we would crank, and the centrifical force pushed the honey from the frames, and it flowed out through the spigot.

After all the honey had been extracted from the frames, Mom put a big shallow pan on the stove with a small amount of water in it. Then the frames or combs were placed into the pan and heated up: we were given pieces of wax to chew, and we would draw out the delicious remnants. It was heavenly!

If you have any beeswax it is very good to keep your sewing needles clean and slippery so they pass through material easily. Also, you can draw thread across the wax; it will keep the thread from drying out and it will preserve the thread, as well as helping it to pass through coarse material.

Beehives packed for winter.

Cranberries Meant So Much

by
David Pelton

Growing up I worked on many cranberry bogs after school and on weekends to buy my first car. It was hard work! But I enjoyed every minute of it. The air was clean and the trees and foliage were always stunning in the fall.

The first bog I worked on was at Edaville Railroad. They had acres of bogs that needed harvesting and there was always plenty of work. I will always remember the sights and sounds of the trains as they made their way through five miles of tracks around the bogs as we worked. I also remember on Fridays we would go to the Atwood home and Mrs. Atwood, the wife of Ellis D. Atwood, the founder of Edaville, would always offer us a cold drink and a big smile as she handed out our pay in a brown envelope. She was a remarkable woman! I also met a lot of good friends while working on the bogs.

My dad was John Pelton, who passed away in 1988. He worked for Ocean Spray for twenty years. His job in the fall was to check in the trucks and inspect the berries as they came from the bogs to be processed. In harvest season my dad worked seven days a week and twelve-hour days to provide for our family. He loved his job and was well known and respected by his co-workers and the cranberry growers.

My dad would get out of work on Friday nights when working in the fall and go to the grocery store and buy bags of food and treats and drop them off at the Lakeville hospital. He would never give his name; he would just say, "Pass these out to the kids." He was a good man, and a great dad.

My wife and mom worked at Ocean Spray in the fall as screeners. I can remember coming home from work and my wife would just be leaving for her job at Ocean Spray to help support our two children. My mom also worked there for many years. During the holidays we always enjoyed the best of cranberry recipes made by my wife and my mom.

I now live in Myrtle Beach, South Carolina, with my wife and son. It's a long way from where I grew up. But I will never forget the times and memories we had, and how cranberries were so important, and meant so much to my family.

Pa's Bogs

by
Alan Centeio

My grandfather was Manuel Centeio. We used to call him Pa. Along with his brothers, he owned cranberry bogs in south Middleborough in the 1950's and early 60's. When we were little, my brothers, sister, and I would play around the bogs while Pa worked.

When I was between ten and twelve, we kids would help Pa harvest the berries. We would scoop them and put them in wooden boxes, which we would then load onto the jalopies. Cars would stop along Route 28 and people would get out to take pictures of us scooping. I remember feeling like a celebrity.

On hot days we would eat ice cream. Pa and Ma owned an ice cream shop called the Dairy Whip, which was located right across from the bogs. They also sold hamburgers and hot dogs. I also remember a nearby fresh water spring that he would take me to. Water would be shooting up like a water fountain. I can recall that clear, cold water touching my lips.

One time, one of the Kennedy boys and one of his cousins came to Pa's bogs. Pa also worked as a grounds keeper for their Hyannisport compound. We boys were all between twelve and fourteen years old, and we went to the bogs to look for frogs and snakes. My parents had gone out somewhere and my mom hoped to get back in time to take pictures, but didn't. I was just happy to have someone else who liked to look for frogs and snakes.

Manuel "Pa" Centeio

In the winter we would skate on the bogs. I learned to skate there; I remember using a cranberry box to keep my balance. All the neighborhood kids would come to skate or play hockey. We would be at it all day. At dinner time, my mother would stand outside and ring a bell to call us in. After we ate, we'd rush outside again. At

Dairy Whip, 1952

night we would build a bonfire and sit around toasting marshmallows and drinking hot chocolate.

It was a good time. I also learned to drive around the bog roads. I have a lot of fond memories, growing up and playing and working on the bogs.

Just a few weeks ago, my daughter saw one of the old scoops and asked what it was. I told her, "That's what we used to scoop cranberries."

It is so much more than just a cranberry scoop. It's a part of my history as well as hers and her brothers. Someday, they'll understand that history, and that behind every glass of cranberry juice or cranberry sauce is someone who grew and cared for the cranberries. They scooped them up, loaded them and turned them into something else for all to enjoy.

Sanding, Digging, and Dusting: A Photo Essay

by

John B. Howes

STURDY
MOTOR
OIL

A Cranberry Insight

by
Maria Green

Working in the wonderful world of hospitality, in the middle of many tourist attractions, I have found myself explaining "The Red Sea of Floating Cranberries" to many a traveler.

One of my earliest memories of "The Cranberry" takes me back many years. On a hot summer day my family would fill up the old station wagon (to over capacity), then flip a coin to see who would be the lucky ones having the first turn to sit on the lowered tailgate as soon as the car hit the dirt road. That dirt road would always lead us to our favorite pond. Of course we would take the long way around the Makepeace bogs to get there. But boy oh boy, it was worth it.

If I close my eyes I can still see the clouds of dirt behind us and hear the rocks hitting the bottom of the car, as we drive oh so slowly along the narrow roads between the bogs. At times the car would get stuck; everyone would have to pile out and form a row of little ones to one side, out of harm's way, as the older ones pushed the car through the soft sand, until everyone got the okay to pile back in.

Every autumn we would scoop up the cranberries the bog workers left behind and make the most delicious cranberry muffins I can remember, and get a newcomer to try a raw berry and let them find out for themselves just how bitter a fresh cranberry can really be.

Driving along the back roads in Carver during the cranberry harvest a few years back I came across some bog workers taking their lunch break. I jokingly asked them if I could get some cranberries. One of the workers jumped onto the bog and scooped up some berries with his bare hands, put them in a brown lunch bag and gave them to me.

The "Cranberry Scoop" image is now on display in my home in a shadowbox that contains the dried cranberries scooped up by that bog worker on that day. The "Cranberry Scoop" story has become one of my favorite cranberry memories.

The Brothers of the Bog

by
Anthony Gabriesheski

It was Thanksgiving time in Massachusetts, a lovely and appropriate time for a drive through the county of Plymouth. This is what we felt as my lady friend and I drove past a cranberry bog in Duxbury on a fine November day. I had driven past this bog many times over the years, yet had never seen anyone tending the bog or tending the red shack set up at the front.

The shack was small, of a kind that seemed to be designed for the selling of the fruits of the bog, complete with a space in front to accommodate a car or two.

As we came upon the location I was surprised to see the door of the shack open and several benches set out with what looked like small boxes of red cranberries. There was no "cranberries for sale" sign or any other indication that this was now a place for their purchase. However, due to the singular nature of this occurrence I suggested to my companion that we stop. She agreed and we pulled over.

As my car came to a stop two men emerged from the dark interior of the shed. Denim overalls with straps, along with three-day beard stubble made them look somewhat unkempt.

Although separated in age by a few years, facial features gave the impression that these individuals were closely related. We noticed several small signs on the boxes indicating the price, and said hello to the men, who just nodded.

My friend looked at me and I nodded too, encouraging her to buy a box of the beautiful red berries. Money was exchanged, the berries handed over, and the men disappeared into the inky blackness of the weathered red shed, all without exchanging a single word.

As we pulled away we both looked back, she out her window and I through my mirror, hoping for another glance of the curious men. We were disappointed with no further appearance, but vowed to come back along that route later in the day.

As it happened we were not gone long, perhaps an hour. We passed by the red shack on the return trip. No benches laden with berries were visible and no trace of occupancy was discernible. Shutters had been closed over the door openings, as I had seen them fixed in the past.

The berries proved to be delicious, and were a big hit at the Thanksgiving feast. Everyone was fascinated by the fact that the berries were local and bought at a bog. As I thought about our strange purchase experience something occurred to me. No vehicle was present at the bog when we pulled up and no other tire tracks were evident. How did those two men get there, and how did they leave?

Invention

by
Ruth M. Griswold

My father-in-law, Gordon Griswold, worked for George Briggs in Manomet after World War II and invented a way of harvesting cranberries after the bogs were flooded. He took an airplane engine and attached it to a flat-bottomed boat. This was used to stir up the ripe berries that would then float to the top, to be raked and gathered up.

Harvest Time

by
Joseph Wilbur

In the early 1980's I was just learning to drive tractor trailer. Picking up cranberries in the fall was a great time. We would use 48-foot trailers that were made to haul oranges; they had high sides and were open at the top.

Finding some of the bogs can be a challenge. They are off roads you travel. That dirt road you drive by looks like it winds deep into the woods. Swing the truck wide onto this road, under the tall cathedral pines. Branches bounce off the top of the truck. Your heart quickens, hoping you've turned down the right road.

The road opens up into a vast meadow. You're greeted by the sight of a red sea of floating cranberries surrounded by a blaze of autumn colors. You pull up alongside the flooded bog. A conveyor is used to load the berries into the trailer.

The growers are always glad to see you. They worry about an early heavy frost. The smaller growers have a lot of family who pitch in with the harvest, and for the family member that wasn't paying attention or was goofing off, a cranberry turned projectile was headed his way. Another funny sight was to see

who would be the first one to step into a ditch or hole and have his waders fill with water.

Now it's time to leave. Your truck is full and the trailer is heavy with cranberries, bits of leaves, and saturated with water. You run through in your mind what roads you're going to take back to the plant; how wide are they? Do they have sharp left or right turns at the intersections or hard corners? Will they take you through busy residential areas?

Okay, ease the truck onto the main road. The trailer creaks and groans; the truck strains under the heavy weight. Keep it at a safe speed. Water runs off the trailer. It will leave a trail almost all the way back along with a parade of cars behind you.

Back at the plant drive onto the scale to be weighed. See that long narrow metal platform? Back the truck and trailer straight onto it. Set the brakes and get out. Two massive hydraulic pistons

slowly rise out from underneath; they lift your truck and trailer almost straight up into the sky.

Cranberries tumble and roll out of the trailer into a big tank of water. Washed and separated, they will continue their journey into the processing plant.

Helping Hands

by
Edward Lodi

When I was a teenager working on the bogs in the 1950's, I enjoyed listening to the older men and women talk about what life had been like during the Great Depression and World War II. One of the more interesting facts I learned was that one year Jamaicans had been brought in to harvest the cranberry crop. No one, however, seemed able to tell me much about the circumstances, or why these exotic workers were imported into Cape Cod and Plymouth County.

I now understand the reason why my co-workers did not have much information to impart. The year they recalled was 1944; most of them, men and women alike, had been away from the bogs, either serving in the armed forces, or otherwise involved in the war effort, making munitions or medicines or building ships. In fact, their absence was the reason why the Jamaicans had been invited to southeastern Massachusetts in the first place.

In 1943 there was such a labor shortage that for a while it was proposed that German and Italian prisoners of war be employed on the bogs to do the harvesting and other necessary work. For whatever reason this never came to pass. I suspect that the logistics of transporting war prisoners to and from isolated locations—at a time when gasoline was being severely rationed—and guarding them would not have proved practicable.

Instead, in 1943 men from Kentucky were brought in. As reported in the Friday, September 3 issue of the *Middleboro Gazette*: "These are all white, independent small farmers. Severe drought for a month has hit their corn and tobacco crops, and the crops have been or are being cut a month early, so that the farmers will be free to help with the cranberry harvest. The workers will be youths of 16 to 18 and men from 45 to 60 with a few of draft age but ineligible for service for various reasons."

Individual growers provided housing for some of the Kentuckians; others stayed at the King Philip Inn in Lakeville. For the most part the workers proved satisfactory, although the *Gazette*, in its September 24 issue, reported that three of the young men, one aged 18, the other two juveniles, were brought into Fourth District court on a charge of disturbing the peace; it seems they had had a bit too much to drink. Two others were sent back to Kentucky because they were "terribly homesick."

All in all, though, the venture was a success, to the extent that some of the workers remained in Massachusetts to work throughout the winter at Cranberry Canners, Inc, in Hanson.

However, in the following year, 1944, these farmers were once again needed to harvest crops in their home state of Kentucky and were no longer available to help with harvesting and processing cranberries. It was then that the War Food Administration and the Extension Service of the Department of Agriculture turned to the British colony of Jamaica.

For those not familiar with the many privations brought on by World War II, it may come as a surprise to learn that one of the obstacles faced by the government was the lack of ships available to transport the workers from their Caribbean island to New England. In the August 25 issue of that year the *Middleboro Gazette* reported: "Two hundred more Jamaicans are due here by September 1 to join with the 150 who arrived earlier...but there is delay...in getting a boat to bring them here."

The Jamaicans proved to be excellent workers. According to Miss Sue Pitman of the Cranberry Sales office, in referring to the earlier group, "They are doing a beautiful job of weeding the bogs. They are ready and waiting to go to the bogs when they are called for and they want to work more than eight hours. They are more dependable than some of the workers employed locally, who lay off if they happen not to want to go to work. The men are well educated, have at least a high school education, are courteous and respectful, and have made no trouble whatsoever."

This first group had arrived in Plymouth on June 18 after a week's voyage (paid for by the War Food Administration) and were housed at South Street in Plymouth at Camp Manuel, an old army camp, where they had the use of seven barracks, two mess halls, a recreation hall, and an infirmary. The men each paid one dollar a day for room and board.

They ate well. A typical lunch, which they brought with them to the bogs, consisted of up to four large sandwiches, a quart of milk, an orange, and a piece of cake. Some of the growers occasionally provided additional treats, such as ice cream or watermelon. According to an article in the *Boston Globe*, the

Plymouth High School dietitian, Miss Helen Dougherty, purchased the food and served as stewardess for the camp. "She plans meals with a thought to vitamins, proteins and calories."

A Plymouth nurse, Miss Gladys Leland, was in charge of the infirmary. Other than bad colds, which some of them suffered the first week, possibly as a result of the drastic change in climate, the men remained in good health.

Eighteen of the men were employed by Ellis D. Atwood, who could not praise them enough. "They are splendid workmen," he was quoted as saying. "And it's a comfort to know that we don't have to depend on weekend labor to harvest the crops this year. Notice how beautifully they swing their scythes—yet none of them ever had handled one. They use machetes, I believe, in Jamaica."

Apparently the Jamaicans enjoyed their experience of working on the cranberry bogs. Their one complaint seems to have been that, although they enjoyed playing soccer on their time off, they missed—because of the lack of proper equipment—playing their usual game of cricket.

Wind Machines, Newspapers, and Algal Scum

by
John B. Howes
(circa 1940)

Wind machines are also used to protect bogs from frost. These are usually motor driven airplane propellers set up on a revolving platform. These are started up when the temperature begins to reach the danger point. The propeller sets up a circulation of air, pushing out the cold air and drawing in warm air. For this reason they are effective only if there is warmer air in the vicinity above the machine or a fire in front of it.

Smudge fires have been used to drive frosts away by burning damp vines all along the edge of the bog. I also know of one case where a small piece of bog was covered entirely with newspapers one frosty night and successfully, too, though the whole countryside was covered with papers the next day.

Algal scum often develops in the flood water when it is held late. This sometimes dries to form paper over the vines after the water is let off and is then harmful. It can be prevented from forming by dissolving four pounds of copper sulfate to the acre-foot in the flowage about the first of April. Coarse crystals of the chemical in a burlap sack may be towed in the water behind a boat.

A spray made by dissolving seventy-five to one hundred pounds of salt to one hundred gallons of water is used to control Fire Weed, Wild Bean (Ground Nut), and Partridge Pea. This spray will not harm blossoms, vines, or fruit. Many growers substitute sea water.

Water-white kerosene has rapidly come to the front as a spray on bog weeds. Three hundred to six hundred gallons an acre of it applied as a spray or with a watering pot, the second week in May, is a good control for grasses, sedges, rushes, loosestrife, alders, and brambles.

Frequent use of the insect net will inform you of the development of insects on your bog. If fifty sweeps of a net eleven inches in diameter gather more than nine gypsy caterpillars, nine cutworms, thirty-six spanworms, or three blunt-nosed leafhoppers, treatment is necessary. Combinations of these worms may be figured on the basis of four spanworms equaling in damage one cutworm or gypsy caterpillar.

The net should be made especially for bog work, and its mouth should be about eleven inches across. Such nets are made under the supervision of the Cranberry Station at East Wareham, and may be purchased there at cost.

If it were given to me to select the leading cranberry grower of the country, I would name Mr. Clayton McFarlin of South Carver. I would not name him because he has extensive cranberry properties, for his holdings are only moderate. I would not name him because he gets a higher average yield per acre than others, though I think he may do this. I would not name him because he gets a higher average net money return per acre than others, for I know nothing about that and doubt if it is true. I would not name him because I think other growers should very generally adopt his methods, for that might be impractical. I would name him because more than anyone else he has made of cranberry growing a fine art.

A well kept cranberry bog is always beautiful, especially when in bloom, but when one visits Mr. McFarlin's bogs, and comes away, he finds within him a desire to go back and see them again.

Wakeup Call

by
Winston F. Bolton

I had spent two years in Japan with MacArthur's occupation forces prior to my discharge the previous September, and had been hanging around the house since, conspicuously unmotivated. "How'd you like to rake a few cranberries?" my father asked. This came from the top of the staging on the west peak of our house in South Duxbury.

I looked up from the grass, where for the past half hour I'd been studying an ant laboring among Walt Whitman's blades of green. "Sure," I said, thinking the moment would pass with his brisk stroking brush.

"Mr. Delano's looking"...He paused to stroke in some more paint on the dry shingles, catching a missed spot. "Looking for...damn this Latex!...for some help with getting the harvest in."

My father was more one to *suggest* than force an issue. But I was beginning to get the hint by now. "Sure, Dad. When?" I heard the brush slapping harder. The next afternoon I drove over to the Delano bog, off Route 3A in the north part of town. I was greeted by Mr. Delano himself, who seemed pleased to have me there; I had gone to school with Phil and Martin Delano. He said he'd spoken to my father—wondered what I'd been up to since leaving the service. "Tells me you're writing a novel." I felt my face reddening. A novel? I hadn't been writing anything, just practicing my Japanese script. In fact, I had outdone Hemmingway's "Soldier's Home," for the past year. It wasn't laziness, I told myself; it was the war. A war I'd had no combat role in, except for one of high anticipation, when we were all out digging slit trenches between the barracks, expecting MIGs over our cantonment area at Itazuke Air Base. But that hadn't happened, and all I ever saw of Korea was the name on a map. Unlike Hem, I'd gotten no shrapnel in my legs, no Purple Heart, or embrace of an ardent nurse. My wounds were embarrassingly restricted to a couple of pulled teeth before leaving Travis Air Force Base for home.

It had been exceptionally hot that fall, as if July and August had no intention of letting go. I was already beginning to feel the sun's blaze on the back of my neck and head when Mr. Delano handed me a small cranberry scoop with no further instructions than, "Go out and meet the others." Too late to escape, I waded into that stifling, humid, airless bog, with a hostile sun shooting its arrows at me. Most of the work had already been done, as I observed. The men, twenty or so Cape Verdean farm workers, were swigging water from a jug handed around. Although farm workers were not strange to my town, none of them looked familiar. One laughed when he saw my scoop; theirs were all nearly twice the size of mine, which to them must have looked like a toy.

"We're workin' the sides of the bog now," an old man told me. "Here—you ain't used to doin' this, are you?" He showed me how to scoot down and tackle the berry vines with the tines of my wooden scoop. "Jus' dump the berries in the barrel."

I was having about as much luck raking those berries as I'd had the few times I'd taken up a pool cue to place a ball in a pocket. The berries would not easily let go, and as I labored into the afternoon, they seemed more and more to be intentionally clinging to their vines. My knees hurt; I was kneeling most of the time, sweat pouring into my eyes. They were passing around the water again, and water never tasted better, though the mouths of twenty others had been on the same "nipple" before mine.

I actually managed to finish the day—masculine pride will take a lot of punishment. They'd have had to carry my martyred corpse out of the bog before I would let myself wimp out under the amused gaze of those stoic veteran workers. But with that day, ended Bolton's short, unhappy career as cranberry picker. I never looked back.

It took approximately another week of concentrated self-indulgence to banish the aches, and the memory of them, from my resentful muscles. But the experience had been an invaluable lesson in career motivation. Those cranberries had shown me that I'd best end my life of ignominious idleness, get my act together, and move on to college under the generous sponsorship of the G.I. Bill.

At Boston University I studied with poet Robert Lowell, who, as irony would have it, would use my sometime employer's ancestor, a Captain Amasa Delano of Duxbury, Massachusetts, as the principle character in his play "Benito Cereno," in *The Old Glory* collection. Which gets me to wonder about that business of six degrees of separation, and if it can apply to literature as well.

I drove out by the Delano bog toward sunset one day last January and got out to take a look. It was cold, about 20 degrees, a light snow, not much wind, but the temperature dropping; I just stood there a while without my coat on.

Smalley Bog: A Youngster's Paradise

by
Dorothy Jesus-Britto

When I recall one of the most beautiful places I have seen in my lifetime and when I seek peace, contentment, quiet, and comfort, I remember my youth and the natural beauty of Smalley Bog.

My father was foreman-caretaker for a large cranberry grower; our family lived in a house high on a hill that overlooked a few hundred acres of forest, enclosing about a hundred acres of cranberry bogs. Except for the six or eight weeks when migrant workers came with their families to live in the scattered shanties and harvest the berries, ours was the only family in this magnificent domain—the only exception being a few bachelors who inhabited one or two of the shanties year round.

Often in the spring or summer, I found myself wandering off alone to seek solitude in the forest closest to our house. Although there were many paths connecting different sections of bog, I had a favorite which led me to my own chosen space,

where I plopped myself down on a bed of pine needles. Thick, soft and comfortable, the bed allowed me to lie quietly as I tuned in to the sounds of nature.

Relaxing and finding my inner center, I listened to the sounds of the forest, to hear the buzzing of insects above my head and the chattering of squirrels and chipmunks. Periodically, I saw creatures flying from tree to tree: flying squirrels, whose webbed feet allowed them to maneuver the short distances safely.

I tried to identify the birds by their plumage or songs as they perched on the tree branches and serenaded me. A little stream, running by the side of my spot, allowed me to listen to its music while the water rushed through, over the rocks and stones, into the reservoir close by.

Wild ducks, geese, and other fowls were in abundance. The birds that fascinated me most, however, were the crows. I soon learned that they—much like parrots of the tropics—imitated the sound of human voices. The first time I heard this, I ran excitedly to share this strange phenomenon with my father. He laughed at my amazement but quickly confirmed my findings.

"Most times," he explained, "the crows repeat the sound of words they hear." After that, it was not unusual to hear my father's voice, sounding raucous but clear, screeching from the throats of one or more of these crows.

The huge reservoir held water to feed the bogs. It also kept the bogs from being destroyed by my father's arch enemy, frost, which he fought every spring and fall.

Here was where my siblings and I spent our free time in the summer, swimming, fishing, and boating. It was where we took our boat to search for and pick beautiful white pond lilies. The lilies gave off a very pungent, sweet-smelling aroma and were surrounded by emerald-green pads or leaves. It was very difficult and treacherous to pick them. To get close enough, we often collided with tree stumps protruding from the bottom of the reservoir. But we always managed to collect enough to present to our mother.

Exploring our vast water domain was always a different experience. We were never sure what we might find as we rowed and told each other to be very quiet so that the frogs would not hear us. We believed they croaked for us only if we remained quiet. In reality, of course, they were letting us know we were disturbing their privacy.

Other times, if we sat quietly, we heard the droning of the dragonflies. We called these insects "sewing needles" because someone had told us they could sew our lips together. Not daring to test the strength of the fable when they were too close, we remained very quietly praying that they fly away.

My siblings and I learned to fish in the reservoir, but the grownups had priority there. Since it was abundantly stocked with several varieties, we youngsters found it easier to fish the overflow of perches and hornpouts in the ditches that separated the bogs.

The more experienced adult anglers fished the reservoir exclusively and caught—along with perches and catfish—bass, pickerel, eels, huge snapping turtles, and trout. I don't recall any of us youngsters catching these because it took special skills which we had not learned.

After we brought our catch home, my mother taught us the skills of cleaning fish, after which we had our own fish fry. My mother refused to let us fry fish in her kitchen as she objected to our getting in her way. She did allow us to go down the hill to one of the abandoned shanties and start a fire in the stove. There, we fried the fish in pans we had borrowed from her. All our efforts produced the most exquisitely-tasting fresh fish that we devoured with ravenous appetites.

During harvest, Smalley Bog came alive with many people. The migrant workers came from surrounding towns, while others came from big cities in Massachusetts and other states, bringing their families to live in the shanties for the six to eight weeks' season.

The period of which I write was the depression years, the 30's and early 40's, when people were desperate for jobs. Cranberry harvesting allowed them enough supplemental revenue to feed and clothe their families. Their only other income was government welfare; in some cases, starvation was a constant threat.

We had plenty of playmates during this season. Most of the younger children had been given permission by their schools to join their migrant parents during harvesting, but my parents did not allow us to stay home from school. We were anxious to rush home after school, change clothes, and join our friends on the bogs. It was great fun for us being with others our age.

In the early evenings, some of the children gathered at our house on the hill because my mother did not allow us to visit them, but she did welcome our new-found friends in our home. We often gathered in a group outside our house, lit bonfires to keep mosquitoes away, told stories, jumped competitively in the sand holes scattered around the property, and learned the latest dances from our more sophisticated city friends.

Most of these children came from cities about which we had only heard. They too had interesting stories of their own preconceived beauty and enjoyment. Although foreign to us, these delighted our senses. We, in turn, captivated them as we told of our adventures, excursions, and experiences. After harvesting, when the families had returned to their homes in the cities, Smalley Bog became isolated once again, strictly the domain of our family and the few bachelors who lived year round on the property.

In the winters, my father covered the bogs with water pumped from the reservoir to protect the vines. As a result of the weather, the bogs froze over several times. This freeze allowed us to spend much of our free time skating on the bogs. We were not allowed to skate on the reservoir because of its depth. A constant was my parents' fears of our drowning if we plunged through the thinnest part of the ice. The bogs were always safe, for we did occasionally fall through the ice, but the deepest we submerged was waist or chest high.

Sometimes, after a severe snowstorm, we were not able to go to school. My father waited for the company truck to plow the access from Smalley Bog to Tihonet Village, where Wareham's Street and Road Department plowed its paved roads. Staying home from school under these circumstances was always a pleasant surprise for us as my mother crept quietly upstairs to wake us gently and whisper, "Go back to sleep; no school today." We didn't go back to sleep, of course. After a piping hot breakfast, we dressed warmly and went out to play in the deep snow, building snow people and forts for hours.

We had so much land space that my mother designated several plots for us to plant our own gardens in the spring. My father, with the help of my older brother, turned the ground over. My mother taught us how to plant seeds of the previously chosen vegetables. This was a fun learning experience as we cared for our plots, watering and weeding. When the vegetables began to bloom, we had the constant task of keeping the bugs away because they competed for the tasty morsels.

Potatoes in bloom were especially interesting for me to observe because of their incredibly beautiful white blossoms. I suppose, if these blossoms weren't so common, they would be considered as precious as the exotic orchids of the tropics.

We didn't use pesticides in those days. Each one, carrying a little can, picked the bugs off the plants by hand, then emptied them into the ditches. We were very vocal in our complaints of this gruesome and offensive chore until my mother explained the necessity if the crops were to survive.

Waiting impatiently each day, we inspected their development, closely watching their maturity to pluck the tomatoes and cucumbers off their vines. Wiping them off on our pants legs when we found ripened ones, we sank our teeth into those delicious creations and moaned with pleasure. Pulling up carrots and radishes was not a priority.

Harvesting potatoes, not one of our favorite jobs, was time-consuming drudgery. My father showed us how to use the potato digger carefully so as not to bruise them, or we used our hands to dig them up. We filled many boxes that my father and brother, using wheelbarrows, carted off to the cellar for the winter.

Several acres of bogs were surrounded by blueberry bushes. These had originally been planted as windbreaks, protection for the newly planted cranberry vines. In time, as the workers cultivated the bogs, the blueberry bushes also became recipients of their care. In the middle of summer, the blueberries ripened and we spent much of our free time picking them for my mother for pies, dumplings, and other tasty goodies.

In our explorations of Smalley Bog, we discovered the unbelievably sweet wild strawberries. After tasting wild raspberries, we spent considerable time arguing and comparing size, sweetness, and texture of each red and purple variety. Along the dikes the elderberry bushes grew. We didn't care to eat them for they were not very sweet, but my mother assigned us periodically to pick quarts for her wine making. She also took advantage of the abundant wild grapes—green and purple varieties—to include in her wines.

Another variety of berry we discovered was the teaberry, pink and very small, that grew along the road from which we walked home as we stopped occasionally to sample crab apples at the same time.

My memories may seem endless, so I'll conclude with these words:

Spring was when I sat on a window box in a house high on a hill, and marveled at acres of trees and bogs bursting with new buds. Summer was acres of pink blossoms covering the entire ground. Autumn was an array of reds, greens, oranges, yellows, and purples as the trees and bogs changed colors. And winter was a full moon illuminating a blanket of snow, covering all the grounds, with shiny frozen, snow-laden tree branches genuflecting in homage to nature, its glory, beauty, and peace.

This was Smalley Bog, a youngster's paradise.

Appendix: An Almanac of Things to Do

Mr. Russell A. Trufant of Middleborough, Massachusetts, a cranberry grower, has prepared yearly Cranberry Almanacs in which he lists the seasonal operations on cranberry bogs as a reminder to growers. In 1941 the New England Cranberry Sales Company recognized the potential value of the Almanac and arranged with Mr. Trufant for the publishing of it as an added service to members. — from Bogs: Cranberry Growing in Massachusetts, *an unpublished book by John B. Howes*

January, February, and March

Cutting logs for shipping boxes for future crops.

Cutting wood for fuel.

Clearing swamps to be built into bogs.

Clearing around the bog and sand pits.

Burning brush and grass.

Nailing up next year's shipping boxes.

Repairing, overhauling, and painting of machinery, tools, and buildings.

Repairing picking boxes, cleaning separators and screenhouse.

Sanding the bog on the ice.

Carting sand to be used in spring sanding.

Inspection of dikes and flumes at least twice a year, stopping leaks, etc.

Creosote the Gypsy egg clusters on the trees around the bog.

April

Draw off water on bogs April first two years out of three, provided you can re-flow.

Sand on bogs this month.

Draw water early this month for later re-flow for White Grub, losing crop.

Change water between the 1st and 15th, exposing the bog a week to ten days, to control spring scum.

Finish up building new bogs.

If you sanded on the ice, smooth up the application as soon as the water is drawn off.

Pull such weeds as poison ivy, Horse-brier, Hardhack, Maples, Chokeberry, Leatherleaf, Sheep Laurel, Brake, Sweet Pepper Bush, Rushes, etc., before the bog dries out.

Get ready for frosts.

Repair flumes and pumps.

Arrange for frost warnings.

Prepare and place your birdhouses.

May

Read the Weed Control Chart and Insect and Disease Control Chart.

Kerosene and pull weeds.

Plan to finish sanding early in month.

Drain your bogs about the 5th.

Re-flow the 12th for White and Root Grubs.

Active frost season begins.

Watch for insect infestations.

Begin regular net sweepings.

Watch the Gypsies. (First spray if needed between the 15th and 20th.)

Pull winter water not yet drawn off about the 20th to 25th.

Flood for Gypsies about the 25th. (Take off rubbish.)

Set vines this month.

Inspect beehives.

June

Begin five 36-hour flows at 10-day intervals on new bogs on the first.

Watch the Gypsies. Spray, dust, or kerosene ditches if needed.

Keep the net going. (Watch out for insect troubles.)

Get fertilizer on before bloom.

Treat weeds.

Watch for frosts.

Be ready to dust leafhoppers if necessary.

July

Treat Wild Bean, and other weeds.

Watch for insects. (Use the net; dust or spray if necessary.)

Drain Grub flow the 10th.

Mow on bogs and dikes.

Irrigate only when necessary.

August

Treat weeds, etc.

Mow on bogs, dikes, and shores.

Irrigate when necessary.

Growers meetings. (Crop estimates.)

Arrange for frost warnings.

Get ready for picking.

Repair pumps and flumes.

September

Picking time.

Watch for frosts.

Pick bogs most liable to frost damage first if possible.

Screen and ship "first-round" berries.

Flood seventeen days for Fruitworm and five days for Girdler before October 1st.

Take off floats
and rubbish.

October

Finish picking.
Flood for floats.
Remove all floating trash from bogs.
Prepare for sanding.
Screening.
Rake vines.
Clean ditches.

November

Sanding.
Repairs to flumes and dikes.
Trap muskrats.
Raking.
Screening, packing, and shipping season.
Be ready to flow new bogs soon after Thanksgiving.

December

Finish sanding.
Winter flood put on this month.
Finish screening.